FISHING
FROM THE
HUMBER

FISHING
FROM THE
HUMBER

Arthur G. Credland

Acknowledgements

This volume and all the royalties are dedicated to the Hull Maritime Society, which has supported the activities of the Hull Maritime Museum since its foundation in 1975.

Almost all the images are from the archive of the Museum; principally the Barnard, Cartlidge and Fussey collections. Mike Thompson kindly provided prints of several of the vessels lost at sea, Lynne Petersen provided a photograph of the *Irrawaddy*, and Pauline Greaves has provided a typescript from the hieroglyphics of my manuscript.

First published 2003

Reprinted in 2014 by
The History Press
The Mill, Brimscombe Port,
Stroud, Gloucestershire, GL5 2QG
www.thehistorypress.co.uk

British Library Cataloguing in Publication Data.
A catalogue record for this book is available from the British Library.

ISBN 978 0 7524 2813 0

Printed in Great Britain.

Contents

Introduction

The development of fishing as a major industry in Hull began in the middle of the nineteenth century. Until then the fish landed at Hull, whether by fishermen working out of the Humber or from other ports, was essentially for local consumption. Depletion of fish stocks in the English Channel and southern North Sea encouraged the Brixham and Ramsgate men to fish further north. They found a ready market in the summer season at Scarborough, a popular seaside resort, and other boats landed their catch at Hull where the newly established (1840) rail link to Leeds meant that fish could be transported inland while still reasonably fresh. The 'Silver Pits', a prolific source of sole discovered in the 1840s between the Dogger Bank and the Humber, further encouraged the fishermen to move north on a permanent basis. In 1845 the total Hull fleet consisted of twenty-one sailing smacks which, by 1872, had burgeoned to a massive 313.

By the 1870s the single-masted smacks had enlarged to 60-80 tons and had a mizzen mast to give a ketch rig or 'dandy rig' as the fishermen called it. To meet the demand for cheap protein from the rapidly expanding metropolis and the industrial towns of the North and Midlands, the box or fleeting system was introduced to Hull by John Sims. Pioneered by the Barking fishermen on the Thames, this involved the smacks staying at sea several weeks at a stretch but transferring their catch each morning to a fast cutter which then rushed it to market in Hull or London's Billingsgate. Ice was first introduced aboard the cutters to improve the quality of fish; originally this ice came from rural ponds stored in an ice-house, but then the much cleaner ice collected in Norway was used and finally in 1891 Hull's first ice factory opened for business on the dockside.

The fleeting system became known as 'Sims railway' and this 'industrialised' approach to fishing yielded a ready supply of fish, which encouraged the development of the fish and chip shop soon to be found in every neighbourhood in the land.

The burden of work on the crew, generally just five (three men and two apprentice boys) was immense. They handled the rigging, shot the trawl and hauled it in, boxed the fish and rowed it to the cutter every morning before commencing fishing again. The sheer drudgery and a monotonous diet of fish and potatoes made the fishermen pray to the Dutch and Belgian 'copers' selling cheap spirits. Incidents of drunkenness and violence became a blot on the North Sea fishery. The death of fourteen-year-old William Papper, murdered by his skipper in 1881 and thrown overboard, led to tighter control of fishing by the Board of Trade and a permanent presence by the mission ships which provided spiritual support, medical help and cheap tobacco and tea as an alternative to the 'copers' evil brews.

Steam cutters were in use by the 1870s and experiments converting steam paddle tugs for fishing were made in Shields and Scarborough. They were a great success and then in 1882 the steam trawler *Zodiac* entered service. Designed and built as a trawler, it was the beginning of a revolution in North Sea fishing; within twenty years the sailing smack had vanished from Hull. Steam power meant that trawling was no longer dependent on the wind and fishing could take place beyond the confines of the North Sea. In 1891 Hull fishermen began to exploit the Iceland grounds and the Faroes became another destination.

The early fishermen had been hindered by an uncooperative Hull Dock Co., initially berthing at the ferry boat dock, then in a corner of the Humber Dock. This resulted in some of the disgruntled smacks-men moving to Grimsby in 1855, encouraged by the Manchester, Sheffield & Lincoln Railway and the Great Northern Railway. It was the start of Grimsby's development as a significant fishing port and as a rival to Hull, on the south bank of the Humber.

In 1869 the construction of the Albert Dock gave fishing craft more space but they still had to compete with the merchant fleet. In 1883 the St Andrews Dock was opened, dedicated solely to the use of the fishing fleet, and such was the demand for space that an extension was built in 1895.

The twentieth century saw the development of new fishing grounds off Norway and in the Barents Sea ('White Sea' to the fishermen). Hellyers pioneered Greenland halibut fishing, though the frozen fish it provided was not accepted with enthusiasm; 'fresh fish' as provided in the now traditional way, preserved in crush ice, was generally preferred. This prejudice remained among merchants and consumers until the modern refrigerated stern trawlers began to dominate the market in the 1970s. Remarkably, a cold air refrigerated plant had been installed in an Aberdeen trawler as early as 1884. A technical success, the equipment was soon abandoned.

During two world wars trawlers were requisitioned and adapted for minesweeping, coastal patrols and convoy work, and many vessels were also built to the basic trawler designs specifically for military duties. The threat of mines, submarines, surface ships and aerial attacks made the North Sea a 'no-go area' and during the Second World War Fleetwood, on the West Coast, became the home of the fishing fleet.

After hostilities ceased, vessels were refitted and reconverted for fishing and new craft began to roll off the slipways in increasing numbers from the late 1940s. So successful was the rebuilding of the fleet that a surplus of fish resulted in falling prices and large amounts of good-quality fish were destined for the fishmeal factory instead of the table.

1961 saw the advent of stern-fishing out of Hull; the net towed behind and hauled up a ramp, instead of from the side – a method which had been used since the Middle Ages. Most of the stern trawlers were also freezers and the fish, when landed, could either be held in cold store or sold straight away for consumption as the market dictated. However, the size of these trawlers and the enormous catching potential of the huge nets, as well as the support of electronic fish-finding equipment, meant that an unnecessary strain was being put on fish stocks. Iceland imposed a fifty-mile limit around its coast which, despite initial antagonism, was quickly accepted. In 1975 this limit was increased to 200 miles – leading to the third Cod War – but this was eventually accepted too. At this time, limits and quotas were also introduced by many other countries. Hull was particularly badly hit as, since the abandonment of North Sea fishing in 1936, it was entirely dependent on the distant water fleet; the port depended on fishing in other people's 'back yards'! The rising cost of fuel as a result of decisions made by the OPEC cartel to restrict production added to the misery and much of the fleet was laid up with nowhere to fish.

In 1964 the fleet, largely consisting of side-fishing trawlers, was some 130-strong, while by 1978 it was nearly half this number. Hull continued as a major fish-processing centre but much of the fish was being landed by foreign vessels or brought over land from other ports. Only two of the old established companies, Boyd Line and Marr's, survived the amalgamations of the 1960s and 1970s and the massive reduction of the fleet after 1975.

Marr's remains as an independent family-run firm but it was announced in November 2002 that the Boyd Line had been sold to an Icelandic company, though the name will be retained and fishing from Hull will continue. Currently one of the two Boyd vessels fishing sails under the Russian flag.

Over the years there were various ventures overseas, including New Zealand, Australia and off the Falklands, but many trawlers were sold off to foreign fishing companies, converted to offshore work in the oil industry or simply scrapped. The current Hull fleet is in single figures.

A slow recovery resulted in some new vessels for the Hull fleet; the *Lancella* and *Thornella*, built for Marr's in 1986, were the first for that company in twelve years. Throughout the world fishermen are confronted by tighter regulations resulting from falling fish stocks. Despite attempts at conservation, major grounds are in crisis and the once prolific Newfoundland cod fishery has been wiped out. Currently the North Sea is in the spotlight and there is no doubt that its critical state would have been reached sooner had it not been for two world wars. The absence of the fishing fleet in what was effectively a moratorium totalling nearly twelve years allowed the numbers of fish to recover. It is salutary to recall that as far back as the 1880s the men of the sailing trawlers were complaining about falling catches! The reduction in stocks which had already taken place was masked by the introduction of steam which increased catching efficiency and allowed the continuing extraction of large amounts of fish. In successive decades improvement in fish-finding, and the sheer size of modern vessels and their nets, has continued to mask ever-decreasing numbers of fish until the sustainability of the stocks of 'white fish', especially cod and haddock, has increasingly been brought into question.

Despite the dark clouds over the industry, October 2001 saw the opening of 'Fishgate', the new Hull fish market, at a cost of £4.5 million. It incorporates electronic selling equipment and the ability to trace every box of fish back to each vessel.

Arthur G. Credland
Hull Maritime Museum

Note: This volume largely describes and illustrates the fishing history of Hull; the story of its rival Humber port, Grimsby, on the south bank, deserves a separate volume.

One

From Sail and Oar
to Steam

Whitby salmon coble *Jennie*. From the Humber to the Tweed, the coble is the classic inshore fishing boat. Clinker-built (with overlapping strakes), powered by oar, sail and latterly engine, it has been used for long lining, laying pots for crab and lobster, and drift netting.

Spurn Point, at the entrance to the Humber, is the home of a permanently manned lifeboat station. At the turn of the twentieth century there was a mixed crew of Yorkshire coble fishermen and Norfolk men with their Sheringham crab boats – shown here with their characteristic ports for the oars.

Two fishermen baiting long lines; coils of fine line with hundreds of hooks each baited with a limpet, mussel or whelk. Buoyed at the upper end the line drops down to the seabed. Still used, this method provides prime fish sold fresh for the table.

Paull shrimpers at Hedon Haven (on the Humber) *c.*1890. Early fishing from the Humber shore was mainly confined to the estuary and until the 1920s significant numbers of shrimpers worked out of Paull (east of Hull) making their catch with beam trawls. During the winter they fished for cod.

Paull shrimper *Venture* (H407) photographed in the 1950s. At 30ft long it was one of the last working shrimpers on the Humber.

Above: Paull fishermen, c.1890, posed for the camera on the Humber bank, with an old hulk in the background.

Left: Brixham fishing smack (registered Dartmouth, Devon) entering Dover harbour, c.1880. For centuries fishing out of Hull and its neighbourhood met largely local needs. In the 1840s the Brixham and Ramsgate men worked off Scarborough in the summer and supplied the demands of the summer visitors, or landed at Hull from where the newly established railway could reach a large market in the industrial north.

Hull-registered smack *Burton Stather* (H583); the true smack was a single-masted vessel but by the 1860s had increased in size and was fitted with a mizzen to create a ketch-rigged (or 'dandy-rigged') vessel. From the 1850s the Devon and Kent fishermen began to settle in Hull.

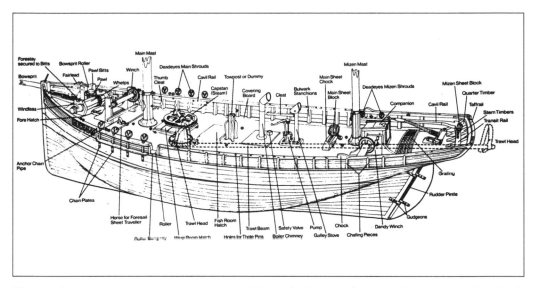

The first hint of steam power came in the 1870s with the introduction of a steam winch, which removed the backbreaking labour of hauling in the trawl net with a hand-operated capstan.

13

The 'dandy-rigged' *Regalia* (H1490) *c.*1885, owned by C.H. Westoby of Hull. The beam trawl can be seen lashed to the port side, an 'iron' at either end which ran along the seabed lifting the net clear of obstructions. It needed a good stiff breeze to provide enough power to tow the net effectively and enable the fish to be retained within the net.

Towing a beam trawl; the maximum length of beam was about 50ft, beyond which it was liable to flex and break, and this of course limited the size of the net.

Above: The shattered remains of the *Young Greg*, 15 May 1896, destroyed when the temporary dam separating the fish dock from the new extension, under construction, gave way allowing a sudden out rush of water.

Right: A mission smack. The early mission crews were themselves fishermen, inspiring the trust of the North Sea men to whom they brought the gospel, medical help, comforts and tobacco. They also made a catch to help defray the cost of the mission work.

The *Spurn* (SH90) at Scarborough in the 1890s, Cuthbert Brodrick's famous Grand Hotel behind and cobles in the foreground. The use of paddle tugs to tow the smacks out to sea when the wind was slack led to the successful conversion of steam tugs for fishing, pioneered at Shields and Scarborough.

The *Patriot* (SH1412) *c.*1890, hauling in the beam trawl.

Precursor (H.1426). Built in 1885 for Robert Hellyer of Hull, a key figure in the development of Hull fishing who had migrated from Brixham. This is the first vessel built by Cook, Welton & Gemmell at their Hull yard on the Humber bank. Though iron-hulled, it was sail-powered only.

The *Irrawaddy* (H1479). The ninth vessel launched by Cook, Welton & Gemmell (in the same year as *Precursor*) this is a fully developed steam trawler. Built for George Beeching and Thomas Kelsall, it measured 100ft long and 20ft wide.

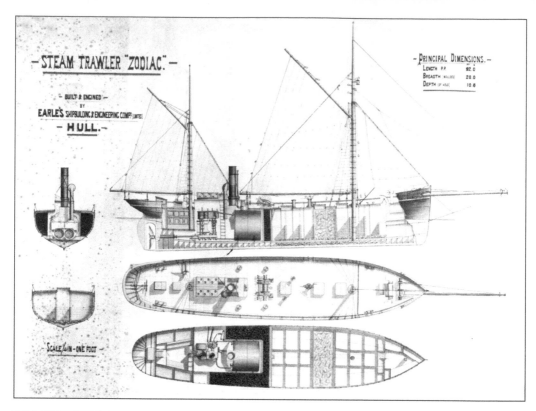

Smack *Leslie* (GY829) of Grimsby. A Priestman oil engine was installed in this wooden smack, seen here in Grimsby docks in 1894.

Opposite above: The *Zodiac* built for the Grimsby & North Sea Fishing Co., 1882, by Earles of Hull. The first built-from-scratch steam trawler, it was 92ft long and 20ft wide. Steam trawlers retained a set of sails many years thereafter because it enabled coal to be economised.

Right: A 100hp marine oil engine, as installed in the smack *Leslie*, built by Priestman of Hull. Their pioneering efforts were overshadowed by the work of Herr Diesel and the firm is best known for grabs and earth-moving equipment.

Below: Smack *Aloan* in the Faroes. Built in 1878 at Dartmouth, she was the Hull smack *John Brown* until her sale to the Faroes in 1896. She was lost in July 1940, probably mined, en route to Aberdeen with a consignment of fish.

William McCann, builder of the smack *City of Edinboro*, 1884, for the smack owners and fish merchants George Bowman and Richard Simpson, at his yard on the River Hull (Garrison side).

The *City of Edinboro*, renamed *William McCann* in honour of the builder, was retrieved from the Faroes and restored in her home port. Nearing completion here in Ruscador's dry dock, Hull, she was in sail again by 1 April 1984, her centenary year. She now belongs to the Excelsior Trust, Lowestoft.

Two

Industrialised Fishing

Silver salver presented to J.R. Ringrose, chairman of the Hull Dock Co., at the opening of St Andrews Dock, Hull, 24 September 1883, and designed for the sole use of the fishing fleet. The North Sea 'box fleet' is depicted in the centre.

John Sims; a fine portrait in oils of the man who introduced the 'box' or 'fleeting' system to Hull, the method being devised by Hewetts of Barking, owners of the 'Short Blue' fleet. The smacks stayed at sea several weeks at a stretch, transferring the catch to a cutter each morning.

John Raspin Ringrose, photographed in 1890. He was a member of an old Hull shipowning family which had mainly traded to Holland.

The smack *Rising Sun*. The five-man crew of a smack, usually three men and two apprentice boys, worked hard handling the sails, shooting and hauling the net and transferring the catch each morning to the cutter, all on a monotonous diet of fish and potatoes. Drunkenness and violence was one of the results of their harsh regime.

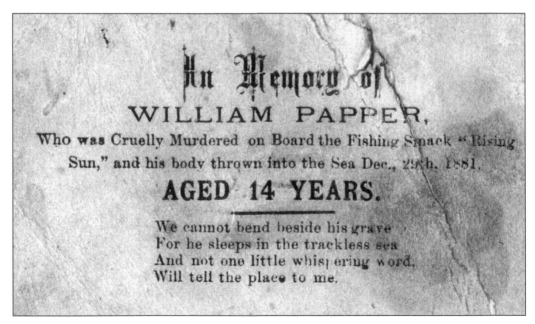

In Memory of

WILLIAM PAPPER,

Who was Cruelly Murdered on Board the Fishing Smack "Rising Sun," and his body thrown into the Sea Dec., 29th, 1881.

AGED 14 YEARS.

We cannot bend beside his grave
For he sleeps in the trackless sea
And not one little whispering word,
Will tell the place to me.

On 29 December 1881, after a series of beatings and general ill-treatment, the boy apprentice William Papper, aged fourteen, was thrown overboard by the skipper Osman Brand and drowned.

The whole story came out after their return to Hull and Skipper Brand and third hand Frederick Rycroft (aged nineteen) were accused of murder. Brand was found guilty and hanged at Armley gaol, Leeds.

MURDER
OF A
HULL FISHER LAD.

Kind friends if you will listen, a sad story I
    will tell.
The murder of a fisher lad of Hull, that has
    befel,
His name was Joseph Rowbottom, 16 years of
    age,
It must have been dreadful to be murdered
    while at sea.
Nicholene and Hadisty must have been of stone
To take the life of that poor boy when he was all
    alone,
No mother to caress him upon the raging sea,
The boy's been cruelly murdered its plain
    enough to see.

Nicholene and Hadisty their wild career has run
Their life they have to forfeit for the crime that
    they have done
Hanging is too good for them with me you will
    agree,
For cruelly murdering that poor boy while out
    upon the sea.
On the 9th. day of January the Sterling went
    to sea,
And every hand upon the deck seemed as
    happy as could be,
The lad was taken sea sick upon the raging
    main,
But now he'll never see his friends or home
    again.

The third hand and the captain cruel men
    must have been,
To kick the lad with their sea boots on, it was
    a dreadful thing,
He asked them to have mercy, when down
    below he cried,
They gave him another shaking and sent him
    stupified.
May God protect these fisher lads when out
    upon the main.
Their task is hard and never know when they
    return again,
They should be treated kindly and well
    protected,
From such villians as these two men either on
    land or sea.

They will soon be brought to justice for the
    crime that they have done
Not a word of sympathy their race has nearly
    run,
They should have shown some mercy to the
    wear lad before,
But now they'll suffer the penalty of the strong
    arm of the law.

Sadly this was not a unique occurrence, and these verses record the death of Joseph Rowbottom, aged sixteen. Papper's death did, however, result in tighter control of the fishing industry by the Board of Trade and the increased presence of the mission vessels on the fishing grounds, working to combat the 'copers' selling cheap spirits.

Begun in sail, the fleeting or box system continued in steam. This dramatic image from the mission magazine *Toilers of the Deep*, 1885, illustrates the difficulty of lowering a boat in the often fierce North Sea weather.

Loaded with boxes of fish, the boat heads towards the steam cutter which will take the catch to market while still fresh; the trawlers continued fishing.

This steam trawler of the Hull Steam Fishing & Ice Co. was part of what was known as the 'Red Cross fleet' because of the company symbol on the funnel. Here a boat returns after unloading its boxes.

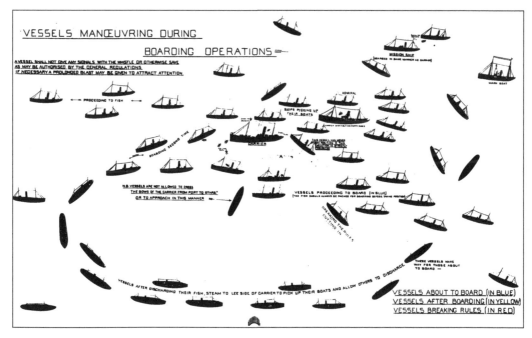

A plan showing the circulation of the trawlers, taking their turn to transfer their catches. These manoeuvres were controlled using flags and rockets from the vessel of the 'Admiral' or 'Don' skipper.

Henry George Foot (born at Brixham, 1844), 'Admiral' of the Gamecock fleet (Kelsall Bros & Beeching) in 1905. Aboard the s.t. *Gamecock*, he is about to take a sounding with the lead line.

'Admiral' Foot, posed with his dog Spot, loading fish boxes over the side into the boat. Boxes are marked Kelsall Bros & Beeching, owners of the Gamecock fleet, named after the cockerel symbol on the trawlers' funnel.

Christopher Pickering (1842-1920) worked in a fish curing house aged ten, became a fish salesman and then began buying smacks and later steamers. Founder of Pickering & Haldane's Steam Trawling Co. in 1888.

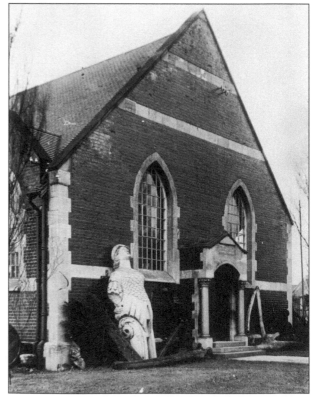

Pickering was a trawler owner, chairman of the Hull Steam Fishing & Ice Co. and the Hull Ice Co. and director of the Hull Steam Trawlers Mutual Insurance & Protecting Co. His benefactions included a park, fishermen's almshouses and Hull's first maritime museum, opened 1912.

Three
North Sea Incident

The box fleet on the North Sea in 1914.

The s.t. *Crane* (H246). On the evening of 21 October 1904 the box fleet fishing near the Dogger Bank came under fire from the Russian Baltic fleet en route to the Far East to engage the Japanese. One trawler, the *Crane*, was sunk and others holed by shells from the quick-firing guns.

Skipper Walter Whelpton, right, leaning on the shell-holed hood of the companionway stairs of the s.t. *Mino*, photographed in the fish dock.

A hole in the deck casing of the s.t. *Moulmein*, pictured with the crew in St Andrews Dock, Hull.

Illustration from a local newspaper recording the removal of the coffin of George H. Smith, skipper of the *Crane*, from his home in Hull.

REMOVING THE COFFIN OF SKIPPER SMITH FROM HIS HOUSE AT HULL.

Opposite above: Funeral cortege of Skipper Smith and William Richard Leggett on the way to the Western Cemetery, Spring Bank, Hull, 27 October, 1904.

Opposite below: Headstone of William Richard Leggett, erected by his mother; behind left is the headstone of Skipper Smith, and right Walter Whelpton who died later from the after effects of the attack.

Below: Commemorative card for the two men killed aboard the trawler *Crane*.

To the   Memory of

THE HULL FISHERMEN,

GEORGE H. SMITH & WM. LEGGOTT,

who lost their Lives through the

Russian Baltic Fleet Blunder,

on the Dogger Bank, on October 21st, 1904.

Hark to the mourners' weeping,
  Sobb'd with 'bated breath.
Whilst in anguish keeping,
  Watch o'er those whose death
Came while perils scorning
  On the mighty deep—
Night—o'er shadow'd morning,
  Marshalling death's long sleep.

List to the indignation,
  From men of every tongue;
The mighty British nation,
  Whose heart's deep chords are wrung.
Hark to the children's crying,
  List to the widow's prayers.
Daughter of Fair Britannia,
  A Nation's grief is theirs.

God His watch is keeping,
  O'er the children's tears,
Daughters of Fair Britannia,
  Calm with love their fears.
Help the lonely widow
  To lock within the breast,
Her sorrow and her anguish,
  Remembering—God knows best.
      —REGINALD T. SHUTTE.

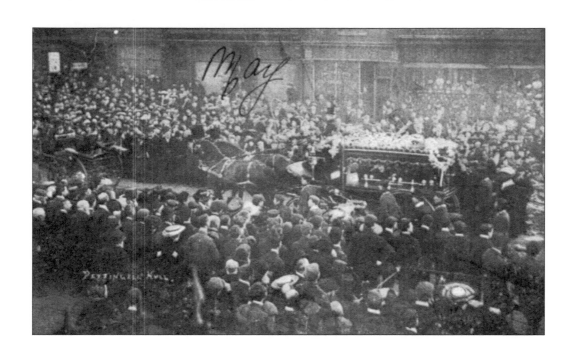

Photographic portrait of Skipper Smith and his wife. (Reproduced from a newspaper copy.)

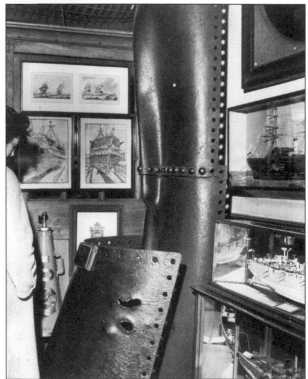

A visitor examining the shell-holed hood of the companion way of the trawler *Mino* at Hull Museum of Fisheries and Shipping, Pickering Park, 1963.

Albert Medal, bronze, second class, awarded to Edwin Costello, bosun of the s.t. *Gull* for his valiant work in rescuing the crew of the *Crane* and recovering the bodies of Skipper Smith and third hand Leggett.

Reverse of Albert Medal showing citation.

Presented by
His Majesty
to
Edwin Costello,
boatswain of the trawler Gull
of Hull, for gallantry in rescuing
the wounded survivors of the crew
of the trawler Crane
when sinking in the North Sea
after damage by the gun fire
of the
Russian Fleet
on the
21-22 October 1904.

Vessel of the Mission to Deep Sea Fishermen, *Joseph and Sarah Miles*, in the Hull fish dock. Further fatalities and serious injuries were mitigated by her presence with the fleet, as she was a steamer equipped with a sick berth and modern equipment, even including an X-ray machine.

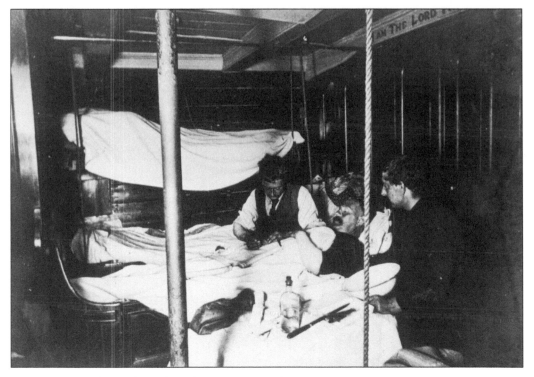

An operation taking place aboard a mission vessel, *c.*1900.

Above: An inquiry into 'the Russian Outrage' took place in Paris in June 1905 and here the fishermen and British officials pose outside the meeting place.

Left: Skipper Fletcher of the trawler *Annapurna* photographed with his wife. He was one of the Hull fishermen attending the court of inquiry in Paris.

Admiral Rojdestvensky, commander of the Baltic squadron. His fleet met the Japanese at Tsu Shima in the Korean strait, 27 May 1905, and was annihilated. The Admiral was wounded and died four years later.

The memorial, a portrait of Skipper Smith in Sicilian marble, was unveiled at the Boulevard, Hessle Road (in the heart of the fishing community) on 30 August 1906. A subscription had been raised by the R.A.O.B., the 'Buffalos', a friendly society popular with fishermen.

41

The box fleet system continued until 1936 when, overnight, while fishing in the North Sea, the company went into liquidation. This is Skipper John Glanville Snr, one of the 'don' skippers in the 1930s.

Four

Trawlers at War

'Armed Trawlers', a watercolour by Frank Mason. Left: *Ameer* of Grimsby, built in 1908, mined 1916; right: *Resono*, also of Grimsby, built at Beverley shipyard in 1913, mined 1915; and centre: *Lord Roberts*, a Hull trawler built at Earles shipyard in 1907 for Pickering & Haldanes, mined 1916.

Left: The *George Cochran*, built to the Admiralty Castle class design at Beverley shipyard, East Yorkshire, 1918; loaned to the US Navy in 1919. Fished for Mills & Co. of London, 1920; sold to Ontario Fishing, 1920, and Newfoundland Transport Co., Nova Scotia, in 1927. Lost with all hands in 1929.

Below: Crew of *Zeppelin L15*, brought down in the Thames on 1 April 1916 by the Purfleet battery, being rescued by the crew of the Hull trawler *Olivine*, skippered by Ernest Elletson. The airship was finished off by 2nd-Lt A. de Bathe Brandon who used incendiary darts. Painting by one of the trawler's crew, J.S. Riches.

Crewman with obligatory gas mask confronted by a cod! Many trawlers, in both world wars, were converted for minesweeping, convoy duties and patrol work, while others carried on fishing, constantly in danger of mines and attack by submarine and aircraft.

Trawler *Darnett Ness* built in 1920 at Beverley shipyard as the *Thomas Boudige* to the Castle class design, one of a number of Admiralty designs which allowed rapid conversion in time of war. Used in the English Channel to tow an electrified cable to explode magnetic mines; she is seen here in the Solent, March 1940.

Grimsby trawler *Aston Villa* (camouflaged with pine trees). Used for anti-submarine work during the Norway Campaign, she was lost on 3 May 1940 after an aerial attack.

The *Kingston Amber* built in 1937 by Smiths Dock of Middlesborough. Seen here on Iceland patrol, she returned to fish for the Kingston Steam Trawling Co. of Hull and was scrapped in 1959.

The *Tehana* (FY525) built by Cochranes of Selby in 1929 for Brand & Curzon of London then sold to Boston Deep Sea Fishing Co. in 1938. After wartime anti-submarine service she was sold to Polish owners and scrapped in 1964.

Chrysea built at Smiths Dock, Middlesborough in 1912 for Sleights of Grimsby. She was used during the Second World War as a balloon vessel for defence of the Humber. Scrapped 1947.

Built at Beverley shipyard in 1944, armed trawler *Homeguard* was a Military class trawler designed for war service. She mounted one 4in gun forward, four 20mm anti-aircraft guns and had a complement of forty men. From 1946 she was named *Loyal* and fished out of Grimsby until 1966.

The Fish class armed trawler *Herring* (T307) on the Humber 14 April 1943. She had a very brief career and was lost on 22 April 1943.

Armed trawler *Inkpen*, a Hills class trawler launched at Beverley in 1941. She mounted one 12-pounder, a 35mm and three 20mm anti-aircraft guns. From 1951 she fished out of Hull as the *Stella Capella* and was scrapped in 1963.

Armed trawler *Hellisay* (Isles class) built 1944 by Cochranes of Selby and seen here on trials, 16 July. Equipped with one 12-pounder, three 20mm anti-aircraft guns and a 40mm gun.

Armed trawler *Imersay* (Isles class) built by Cochranes of Selby in 1944 and seen here on trials in the Humber on 7 December. Equipped as a 'dan layer' for marking out minefields; these are the long poles with floats seen on the upper works, aft. Sold to Greek owners 1948; scrapped 1969.

The *Troday* (Isles class), a dan layer built at Beverley shipyard and completed in March 1945. Post-war she was converted to a tanker in Italy and in 1994 was laid up in Dar-es-Salaam.

The *Azalea*, a Flower class corvette built at Beverley shipyard and launched in November 1939. The design is an enlarged version of a whale-catcher boat and was notorious for lively motion in open water. In 1946 she was converted into a merchant vessel and sank in a collision in 1955.

Lord Inchcape, 2 August 1943, built by Cochranes of Selby in 1924 for Pickering & Haldanes of Hull. Used for minesweeping in the Second World War, she returned to fishing as the *Petten* for Dutch owners and was scrapped 1954-1955.

Lord Beaconsfield, an 'old-timer' built in 1915 by Cochranes of Selby as the *Tribune*. Used as a minesweeper in the Second World War, she is seen here fitted with an acoustic hammer on her bows to explode acoustic mines. Wrecked 17 October 1945.

Another view of HMS *Homeguard*, on the Humber under trial 19 September 1944.

Five

Trawler Builders
of Yorkshire

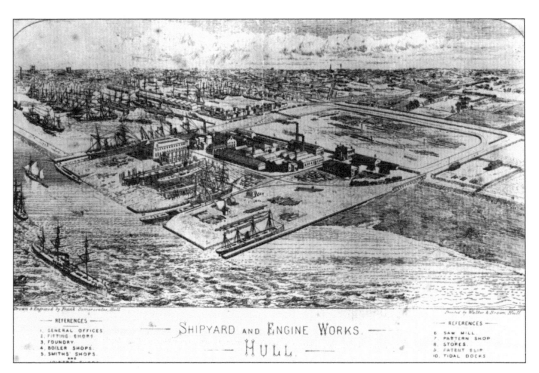

REFERENCES

1. GENERAL OFFICES
2. FITTING SHOPS
3. FOUNDRY
4. BOILER SHOPS
5. SMITHS' SHOPS

SHIPYARD AND ENGINE WORKS. — HULL. —

REFERENCES

6. SAW MILL
7. PATTERN SHOP
8. STORES
9. PATENT SLIP
10. TIDAL DOCKS

Earles shipyard on the Humber bank, on the east side of Hull, pioneered steam-trawler building in 1881-1882 with the launch of the *Zodiac* but after 1908 concentrated on merchant vessels.

The *Australia*, built at Earles shipyard in 1882 as a carrier boat (cutter) for the Hull Steam Fishing & Ice Co., here carrying fish from the North Sea box fleet to London's Billingsgate.

The s.t. *Marshall Oyama* (H836) built in 1905 at Earles shipyard for Pickering & Haldanes of Hull. Note the early form of 'whaleback' at the bows. This helped divert seawater away from the crew's accommodation in the fo'c's'le; the pronounced convexity is flattened in later trawlers.

Opposite above: The *Royallieu* built in 1900 by Cook, Welton & Gemmell on the Humber bank, close to Earles yard, for William Grant of Grimsby and wrecked on the Yorkshire coast in 1906. The first steam trawler, the *Magneta*, was launched in 1885 and the yard's reputation was quickly established.

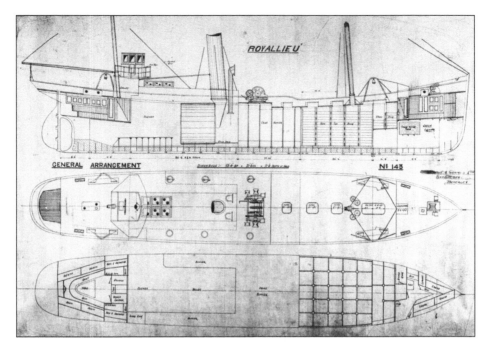

The *Esmeralda*, newly commissioned in 1903 after launch from Cook, Welton & Gemmell's new yard at Beverley where they had moved the previous year. Latterly she fished out of Aberdeen; scrapped 1937.

The *Viola*, built for Hellyers of Hull in 1906 at Beverley shipyard. Sold abroad and converted in 1922 for whaling off the African coast, she was further sold to Cia Argentina de Pesca in 1927 and again in 1960 to Albion Star Ltd, registered at Port Stanley. She was laid up in 1974 and still lies at her moorings in Grytviken, South Georgia.

Bardolph, built 1911 for Hellyers Steam Fishing Co., maintaining the tradition of Shakespearean names used by the company. In 1913 she was one of the first trawlers fitted with wireless and was sunk on 5 June 1915, near the Shetlands.

Opposite below: The *Kopanes* (H502) built in 1930 at Beverley shipyard but here seen refitting on the slips at St Andrews Dock, Hull. Used as a minesweeper in the Second World War she returned to fishing as *Cape Pembroke*, was later renamed *Frobisher* and finally scrapped in Belgium in 1957.

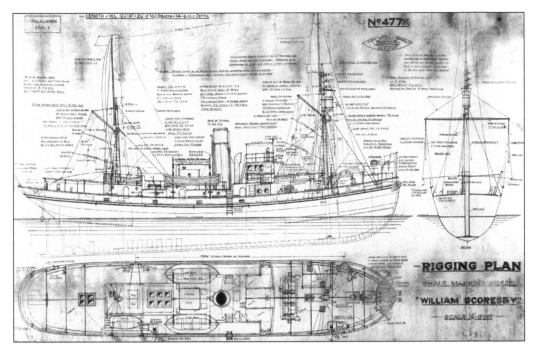

The *William Scoresby*, completed at Beverley in 1926 for the Crown Agents as a research vessel, was used for whale marking and plankton sampling in the South Atlantic then as a minesweeper in the Second World War. She was scrapped in 1954 after a last research cruise in the Indian Ocean.

Lady Lilian (H467) built in 1933 at Beverley for Jutland Amalgamated Trawlers Ltd; converted to Gem class anti-submarine vessel *Jade* and sunk during an air raid on Malta, 10 April 1942, while serving in the 4th Anti-submarine Group.

The *Prince Charles* (H85) was built at Beverley in 1949 for the North Cape Fishing Co. of Hull before becoming *Cape Duner* of Hudson Bros in 1951. Renamed *Ross Duner* in 1960, she was scrapped in Belgium in 1968.

The *Princess Anne*, a motor trawler built at Beverley in 1955 for the St Andrews Steam Fishing Co. Ltd, Hull. She was sold to Fleetwood, renamed *Wyre Gleaner* and scrapped in 1976.

The diesel electric trawler *St Dominic*, built at Beverley in 1958 for Thomas Hamling. Sold for scrap in 1979.

Orders were filled for overseas buyers too, and the *Fylkir* was built at Beverley in 1958 for Fylkir Ltd, Reykjavik, Iceland. She came to Hull in 1966 and was renamed *Ian Fleming* (Newington Trawlers). The vessel was abandoned off north Norway in 1973; three crewmen were lost.

The motor trawler *Stella Leonis*, built in 1960 for Charleson-Smith of Hull, was winner of the Silver Cod trophy in 1963 and 1964 for largest total catch of the year. In 1965 she was renamed *Ross Leonis*, being sold for scrap in 1978.

Above: Skipper Lewis (right) of the *Lord Beatty* receiving the Silver Cod trophy in 1958 for the largest total catch: 40,563 kits valued at £155,903 caught in 330 days at sea.

Right: Steam trawler *Sculcoates* (H51) built in 1924 by Cochranes of Selby for the City Steam Fishing Co. of Hull. Stranded March 1934.

Crestflower, built by Cochranes in 1930 for the Yorkshire Steam Fishing Co. of Hull and lost to enemy action on 19 July 1940, and *Lord Brentford*, built in the same year for Pickering & Haldanes, fitting out in Queens Dock, Hull. After war service the latter went to Cape Town as *Oranjezicht*, being scrapped in 1966.

The Hull fishing industry depended on the skills of many ancillary trades including ship rigging. In 1954 Hollywood came to Hull when the former *Hispaniola* was refitted as the whaler *Pequod* for the film *Moby Dick*. Humber St Andrews provided the expertise and the local museum the historic detail, very appropriate since the Hull museum is mentioned in Melville's book.

Six

Trawler Construction

Aerial view of the Beverley shipyard on the River Hull, some eight miles north of Hull. Note the plates and ribs laid out in the open on the right, two vessels on the stocks and four floating in the river.

Beverley shipyard: inside the fabrication shed, 1958. Note the plate-bending machine in the left background.

Beverley shipyard; trawler on the stocks. The many closely spaced ribs is still very traditional and much like the building of a wooden vessel of generations before.

Beverley shipyard: welder at work, 1955. The speed and effectiveness of welded construction was proved in the Second World War with the mass production of the TID tugs and Liberty ships, which could be assembled in a matter of days from prefabricated sections.

Beverley shipyard: the *Kingston Jacinth*, an oil-fired steam trawler for the Kingston Steam Trawling Co., Hull, entering the River Hull on 3 November 1951. Because of the restricted width of the river, launches were always sideways on. Behind is the *Velia*, launched June of that year for J. Marr.

Beverley shipyard: a dramatic launch with a mighty splash, 10 March 1958, after a fall of snow. The *Northella* was built for J. Marr of Hull, the third of her name.

After towing the empty shell of a vessel round the numerous bends of the River Hull, engines, winches and heavy gear were installed using the sheer legs in Albert Dock. This is *Swanella*, another vessel of the Marr fleet.

Here *Northella* is completing her fitting out at the berth of C.D. Holmes (Engineering) in Princes Dock in the heart of the city. The building with three domes, the former Dock Offices, is now the Hull Maritime Museum.

Most of the trawlers built at Beverley and Selby were engined either by C.D. Holmes or by Amos & Smith (founded 1874). This is the latter's engine erecting shop, Albert Dock, Hull, *c*.1928. The first four sets of engines are the type fitted to distant water trawlers of the time.

Though trawlers dominated the output of the Beverley shipyard, a variety of tugs, naval vessels and specialist craft were also built. In the 1930s this included a series of wooden-hulled minesweepers of the Coniston (Ton) class. *Bronington* (M1115) was depot shop for the Humber Division RNVR 1954-1958 and in 1976 was the first command of HRH Prince Charles.

Aerial view of Cochranes yard at Selby, Yorkshire, 1957 – Beverley's great rival. No less than five vessels are on the stocks and another two recently launched lie in the River Ouse alongside. (The area around the yard has been faded in the original photograph.)

Seven

The Greenland Halibut Fishery

In 1925 Engvar Baldersheim, a Norwegian from Bergen, persuaded Owen Hellyer of Hellyer Bros (Hull) to embark on a large-scale halibut fishery off Greenland. A Dutch-built vessel, the *Helder*, renamed *Arctic Prince*, was converted to carry a refrigeration plant and twenty-eight small boats (dories).

Above: The *Norman* (H249), built 1911 for Hellyers, was one of several trawlers at Greenland from April-October each season working with the *Arctic Prince* recording ice fields and ferrying fish from the scattered dories to the mother ship, which in turn kept the trawlers supplied with coal.

Left: Arctic Queen, originally the *Vasari*, a refrigerated vessel in the New York-Buenos Aires run, was introduced to halibut fishing in 1928. This photograph (29 May 1928) shows one of the dories in the water and the special davits for suspending them.

Above: The *Borodino* (Wilson Line, Hull) built 1911 at Earles shipyard. She and the *Mourino* were chartered to ferry out provisions and any crew replacements and bring back consignments of fish to Hull. Sunk as a blockship at Zeebrugge in 1940.

Right: On board *Borodino* in June 1932 was a German film crew shooting the film *SOS Iceberg*. Also with them was a polar bear brought from a zoo in Europe, since the wild ones were considered too unpredictable!

Left: The star of *SOS Iceberg*, Leni Reifenstal, who was soon to become a film-maker in her own right, notably of *Olympiad*, which captured the events of the notorious 1936 Olympic Games in Berlin.

Below: The *Arctic Queen* under tow in the Humber. The first three seasons produced over a thousand tons of halibut a year.

Opposite above: Arctic Queen in dock. The *Arctic Prince* was used as a refrigerated depot ship in Hull when the season's fishing was over and consignments of frozen fish were released onto the market as required.

Officers of the *Arctic Queen*, merchant seamen from the Wilson Line, Hull. Left to right: Chief Engineer Pattinson; Capt. Thomas; Capt. Phillips; the 1st mate; 2nd mate D.W. Milward; and the 3rd mate.

KONTRAKT

mellem

rederiet og dorybasen med d/s „Helder" (og andre skibe) paa kveitefiske i Davis-strædet (Vest-Grønland).
1928.

Undertegnede Johan Solbu   adr. Garten i Fosna

født 20/8-88 i Ørland   ansættes herved som dorybas paa d/s »Helder for sæsongens kveitefiske i Davis-Strædet (Vest-Grønland).

Dorybasen skal staa direkte under fiskeribasen paa »Helder«. Han godtar ogsaa for sit eget vedkommende i enhver henseende den for doryfiskerne fastsatte kontrakt der er saalydende:

1. Ekspeditionen er beregnet at avgaa omkring 5. mai. Rederiet vil gi fiskerne det nødvendige varsel saa de kan være klar til at fremmøte og gaa ombord til beordret tid i avgangshavnen som vil bli Aalesund.

Fiskerne skal være friske og ved kontraktens underskrift fremlægge lægeattest for at han ikke lider av nogen smitsom sygdom eller andre sygdomme. Ved fremmøte skal han atter la sig undersøke av ekspeditionens læge.

Ved fremmøtet skal fiskerne som personlig utstyr mindst medbringe det som er fastsat i ekspeditionens pakningsliste, og de skal være ombord med sit utstyr til den beordrede tid.

Reiseutgifter i Norge er for fiskernes egen regning. Hvis de avmønstres utenlands betaler rederiet billetten til Bergen.

Rederiet skal bestemme naar hjemreisen fra fiskefeltet skal finde sted, hvilket beregnes at bli omkring slutten av september.

2. Fiskerne skal organiseres i dorylag, der skal bestaa av en dorybas og 4 fiskere. Dorybasen skal selv utvælge sine fiskere, men rederiet skal godta dem før dorybasen endelig hyrer dem. Dorybasen skal være ansvarlig for at de er dygtige linefiskere og i enhver henseende adlyder lederens ordrer, hvad enten de er git direkte eller gjennem andre som han maatte ha ansat dertil.

Hvis nogen i dorylaget gjør sig skyldig i forsømmelser eller ulydighet skal det være dorybasens pligt til øieblikkelig at rapportere det til lederen.

Dorylaget skal drive fisket i nøie overensstemmelse med instruktioner og ordrer fra lederen og ikke efter eget forgodtbefindende om tid og sted.

Saaledes skal det være paa det rene at der ikke maa gjøres nogen som helst forskjel paa fisket og arbeidet paa søn-. og helligdage fra en hverdag.

Den fisker som ikke møter til sit arbeide saaledes som beordret eller ikke avmelder eventuel sygdom før klokken 8 om morgenen skal lederen ha ret til at trække indtil kr. 50.— fra hans fortjeneste. Hvis saadan forsømmelse gjentar sig, eller fiskeren viser sig uskikket til sit arbeide, doven, sniker sig undav, uvillig, opsætsig eller uten tillatelse forlater sit arbeide eller ved ord eller gjerning viser støiende, upassende optræden, skal lederen ha ret til at fratrække indtil kr. 200.— fra hans fortjeneste for hvert tilfælde, og hvis det er nødvendig for at eksekvere dette fratræk, i fornøden utstrækning reducere det træk som er fastsat pr. 15de dag.

Hvis en fisker avskediges paa grund av daarlig opførsel eller paa grund av negtelse av at gjøre beordret arbeide skal lederen ha ret til at sende ham hjem, og hjemreisen skal være for hjemsendtes egen regning. Hjemreisen fra fiskefeltet skal s> e med den baat som lederen beordrer.

Contract of Johan Solbu, one of the dory men. They had the hardest job, spending long periods in their open boats, baiting hand lines and then hauling in the catch, while exposed to the vagaries of the Arctic climate. By 1933 there was a glut of halibut and 1935 was the last season of this venture.

Eight

A Trip

Filling the hold with ice before sailing. Typically a 'side-winder' (side-fishing trawler) spent three weeks on a round trip with two days at home before setting out again.

Ice was introduced initially to improve the quality of fish aboard the North Sea carriers of the box fleet. Natural ice was imported from Norway; one of the vessels being the old Hull whaler *Truelove* here seen at anchor on the Norwegian coast. In 1881 Hull's first ice factory was established.

Departure was decided by the tide; here the crew are going aboard at night, heaving their kit bags over the side.

Right: The Marconi man: though some skippers and mates had wireless qualifications, it was usual to carry a 'sparks' supplied by Marconi who was also able to maintain and repair the equipment. An additional job was looking after the boiling of the cod livers, for which he received a bonus.

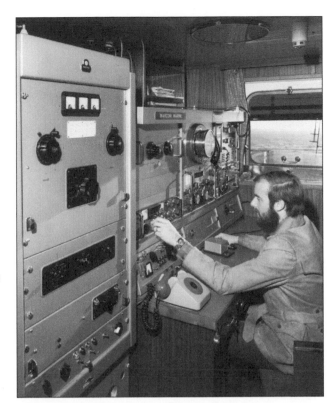

Below: The cook: one of the key men on board – without good food and plenty of it, morale was bad and the crew disenchanted.

Opposite above: Heaving the trawl net over the side; note the line of hollow metal floats. The trawler shoots the net and then tows it until it is reckoned a reasonable catch has been made, the net is then hauled in.

Opposite below: A trawl door hung from a hoop-shaped frame known as the gallows; there was one forward and one aft on both sides of the ship but it was usually the starboard set that was employed. The doors (otter boards) acted like hydroplanes keeping the mouth of the net open.

Right: Cod end of the net hoisted aboard, bulging with fish ready to be released into the deck compartments known as pounds.

Pulling on the cod end rope, tied in a special slipknot, a flood of fish is released onto the deck. Note the large cow hides which help prevent the net being torn on seabed obstructions.

Gutting on the open deck, at night under lights. The cod has been thrown towards the 'washer' in the centre of the deck (off camera).

Pulling the net back on board again to prepare it for the next tow, note the metal hand hook being used.

Left: The fishermen gutted and sorted as quickly as possible. So long as there were fish on deck the men stayed and worked eighteen hours and more at a stretch.

Below: The washer: the gutted fish were thrown into an open metal box in the middle of the deck with a constant flow of seawater through a hose. The natural movements of the vessel cleaned out the fish, which were then sent below.

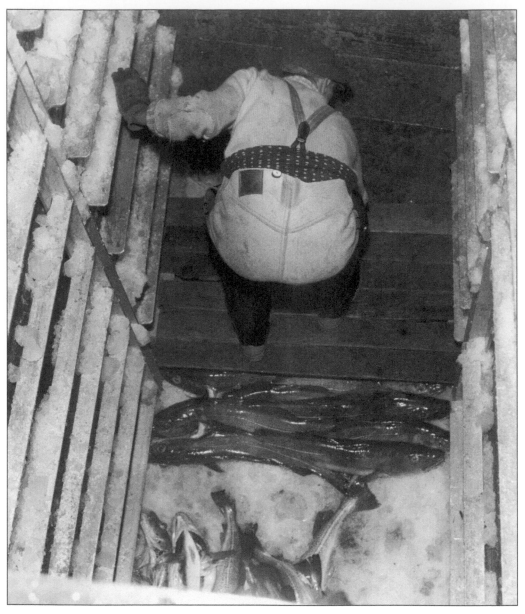

Above: Down in the fish room the catch was stocked – a layer of ice, a layer of fish – until the hold was full. Quality fish was put with a layer of ice on shelves, which prevented crushing.

Opposite above: While their husbands and sons where at sea, dependants claimed a weekly wage or 'allotment' at the company offices. Traditionally pay-day was known as 'white stocking day' because the women would put on their best clothes.

Opposite below: Payment book of Mrs Jenkins, whose husband was aboard the s.t. *Imperialist* 1947-1948; £5 10s 50d a week rising to £6 4s 5d. These amounts were deducted from the final payment made to the fishermen at the end of the voyage, which depended on the value of the catch.

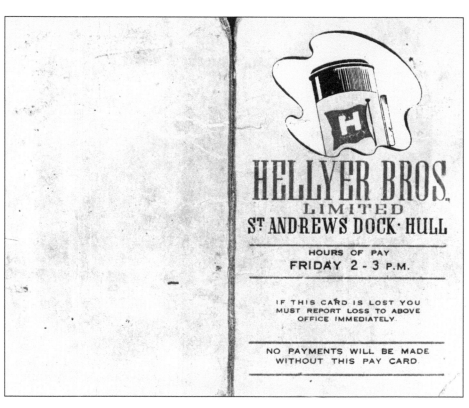

HELLYER BROS.
LIMITED
St ANDREWS DOCK · HULL

HOURS OF PAY
FRIDAY 2 - 3 P.M.

IF THIS CARD IS LOST YOU
MUST REPORT LOSS TO ABOVE
OFFICE IMMEDIATELY.

NO PAYMENTS WILL BE MADE
WITHOUT THIS PAY CARD.

NAME W H Jenkins
RATING 3rd Hand
VESSEL Imperialist

WEEKLY WAGE
PAID TO wife

Date				Date				Date			
30. MAY. 1947	5	10	5	17. OCT. 1947	6	4	5	6. FEB. 1948	6	4	5
6. JUN. 1947	5	10	5	24. OCT. 1947	6	4	5	13. FEB. 1948	6	4	5
20. JUN. 1947	11	.	10	7. NOV. 1947	2	11	5	20. FEB. 1948	6	4	5
4. JUL. 1947	11		10	14. NOV. 1947	6	4	5	5. MAR. 1948	12	8	10
11. JUL. 1947	5	10	5	21. NOV. 1947	6	4	5	12. MAR. 1948	6	4	5
25. JUL. 1947	11	.	10	28. NOV. 1947	6	4	5		6	4	5
1. AUG. 1947	5	10	5	5. DEC. 1947	6	4	5				
15. AUG. 1947	11	.	10	12. DEC. 1947	6	4	5				
22. AUG. 1947	5	10	5	19. DEC. 1947	6	4	5				
12. SEP. 1947	6	4	5	23. DEC. 1947	6	4	5				
19. SEP. 1947	6	4	5	2. JAN. 1948	6	4	5				
26. SEP. 1947	6	4	5	16. JAN. 1948	6	4	5				
3. OCT. 1947	6	4	5	23. JAN. 1948	6	4	5				
10. OCT. 1947	6	4	5	30. JAN. 1948	6	4	5				

Left: Inbetween hauls any damage to the net had to be repaired, sometimes whole sections or even the entire net would have to be replaced after catching on old wrecks and debris on the seabed.

Below: The *Loch Seaforth* (H293) built at Middlesbrough in 1936, berthed at St Andrews fish dock, Hull, post-war. The catch has been unloaded and the boards, which form the pounds, are stacked on deck.

Nine

On Shore:
The Hull Fish Dock

Loading at Hull fish dock, c.1914, dubbed Billingsgate after the fish market in London. The wooden kits in the foreground are stencilled with the name of E.B. Cargill, a Hull trawler owner.

The *Nordborg* (H35) built in 1957. In the 1920s the first Danish seine-netters arrived in Hull. These middle-distance fishing vessels provided a high quality fish sold for premium prices. A fleet was managed latterly by Boyds of Hull but many more operated out of Grimsby.

Hull fish dock, 1936. Bobbers (so-called either because they had been paid a shilling, a 'bob', a shift or from their rapid bobbing movements) unload the catch by hauling baskets of fish from the hold and swinging it ashore.

Above: Not all the fish was sold 'fresh' on the fishmongers slab; up until the Second World War quantities of cod were salted and sun-dried at the so-called 'cod farm' adjacent to the fish dock. Most went abroad to Spain, Portugal and South America.

Right: Andrew Johnson Knudtzon Ltd founded by Anders Jorgensen from Funen, Denmark, who anglicised his name. He was the grandfather of Amy Johnson, the pioneer aviator. AJK still thrives as part of the Marr organisation.

Left: Statue in Prospect Street, Hull, to Amy Johnson (1903-1941) who in May 1930 flew solo from England to Australia in twenty days. Her aircraft was called *Jason*, not after the leader of the Argonauts but from the trademark of the family business.

Below: In 1951 the old-style wooden kits, made with hoop and stave like a barrel, were replaced by aluminium tubs.

Opposite above: Shore-workers admiring a man-sized cod-fish.

Opposite below: Once unloaded and sorted into fish kits (each tub held ten stone of fish) the catch was auctioned, Dutch style, starting high and then downwards until a bid was made. The kit was then given a tally to identify the purchaser.

Edna Aldis, *c*.1935. It was common for wives and mothers of fishermen to make pieces of trawl net at home, hung from a rail, indoors in the cold weather and outdoors in the summer. The pieces were then assembled in the net loft on the fish dock ready for use.

Visiting celebrities often came to the fish auctions and parcels might be sold in aid of charity. Here is the popular vocalist Ronnie Hilton (at the back leaning forward) at the Hull fish market, 12 September 1958.

Filleting shed. Increasingly there was a demand for skinned, boned and ready-prepared fish and the fillet became the housewives preferred buy.

After preparation and boxing, each merchant despatched his fish to his clients around the country. Until the 1960s this was done principally by rail but afterwards it was carried out by lorry and van.

Above: Freshly made blocks of ice in the Hull Ice Factory.

Left: A scene at Hull fish dock in 1930 showing the intense activity and a variety of transport; horse-drawn carts, motor lorries and a railway engine. The fish is stored in wooden kits and boxes.

Right: In the nineteenth century fishermen learned their trade as indentured apprentices but, as more and more 'casuals' were taken on, formal instruction ashore and Board of Trade examinations became the norm.

Below: Huge quantities of herring (mainly imported from Norway) were kippered in Hull and cod and haddock were also smoked. Here two women are packing cured cod fillets.

HM the Queen and Prince Philip, on a visit to Hull fish dock in May 1957, making a call at the medical centre. Dr J. Burns, the medical officer, is centre and H. Watson-Hall, president of the Hull Trawlers Mutual Insurance & Protecting Co. is to his right.

The harbour tug *Wilberforce*, used for manoeuvring vessels in the often congested fish dock.

A tanker barge, *c.*1960, filling the bunkers of a trawler in St Andrews Dock. Post war, most trawlers were converted from coal burners to oil-fired steam engines.

Opposite below: Prince Philip, May 1957, taking a look at the catch as the bobbers unload. In the background is *Kingston Jade* (H.149).

Loch Eriboll (H323) leaving Hull fish dock. Built by Brook Marine of Lowestoft for Loch Fishing Co. of Hull in 1960. Diesel-powered, she was 181ft long and 734 gross tons. She was scrapped in 1979.

Halcyon days! St Andrews Dock, Hull, *c.*1965; the fishing fleet at its peak. A row of side-fishing trawlers in the foreground, *Lord Tedder* on the left. The stern trawler *St Finbarr* is berthed on the right.

Ten

Accidents and Tragedies

Before radar it was not uncommon to run aground, even on familiar coastlines, during fog or heavy weather. Here the s.t. *Ostrich* (built 1891) is temporarily beached near Aldborough, East Yorks., but was refloated. She was eventually sunk by a U-boat in the North Sea in 1917.

On 14 February 1906 the Hull trawler *Southella* ran aground on the south coast of Iceland. The crew scrambled ashore and trekked across the icy, rocky interior before finally reaching Reykjavik. This poor-quality image is the only one available today. Left to right, on ponies: Skipper Thomas Wilson, P. Pelson (engineer) and Christopher Davidson (mate) who received a gallantry medal for his bravery in swimming ashore with a line. Four other trawlers were lost that month on the Icelandic coast.

The *Octavia* collided with another trawler, *George Cousins*, near the Isle of Man, 13 February 1937; all nine crew were saved. The wreck has recently been investigated by divers.

Above: The *Moravia*, built in 1917 at Beverley shipyard and sailing out of Grimsby, was swept by heavy seas off the Westmann Islands in the North Atlantic in April 1932, which removed the funnel, part of the bridge and the mizzen mast.

Opposite below: Two Hull trawlers aground in the Humber in the 1930s. The anchor chain of the *Lady Beryl* (right) had fouled the propeller of the *Alexandrite* (left). Both were repaired and went back to sea.

Two men were lost from the *Moravia*, but these are the survivors back in Grimsby. The boiler fires had been quenched by the seawater and the helpless vessel had to be taken in tow by Skipper Chapman of the Hull trawler *Cape Grisnez*, completing an epic one thousand miles in eight days of fierce weather.

The *Boston Seafire* (H584), an oil-burning steamer, built at Beverley in 1948 for Boston Deep Sea Fisheries Ltd. Renamed *Cape Tarifa* in 1952, *Ross Tarifa* in 1965 and scrapped in 1968.

In December 1951 *Boston Seafire* was hit by a freak wave which smashed the wheelhouse. Taken by a crew member, this poor-quality image is the only one available today.

Colin Neadley (centre) on the wheelhouse of *Boston Seafire*. While recovering from his injuries he began making ship models out of matchsticks and this led to a full-time career as a professional model maker.

Left: The *Swanella* in
April 1960 in Icelandic
waters in sea conditions
which were not unusual.

Below: A 1930s trawler
in heavy weather in the
North Atlantic.

The *Lorella*, built 1947 at Beverley Shipyard for the City Steam Fishing Co., part of J. Marr & Son. She was lost off Iceland 26/27 January 1955, ninety miles N.E. of the North Cape. All twenty crew were drowned.

The *Princess Elizabeth* (H135), built at Beverley shipyard in 1950, she became *Roderigo* of the Hellyer fleet in 1951. Fishing close to *Lorella* she also was lost in heavy seas and a build-up of ice and all twenty crew were drowned.

SAILORS AND FISHERMEN'S BETHEL,
HESSLE ROAD, HULL.

𝔐𝔢𝔪𝔬𝔯𝔦𝔞𝔩 ✠ 𝔖𝔢𝔯𝔳𝔦𝔠𝔢.

SUNDAY, MARCH 19th, 1933, at 6-30.

𝔍𝔫 𝔪𝔢𝔪𝔬𝔯𝔶 𝔬𝔣 𝔪𝔢𝔪𝔟𝔢𝔯𝔰 𝔬𝔣 𝔱𝔥𝔢 ℭ𝔯𝔢𝔴𝔰 𝔬𝔣 𝔱𝔥𝔢 𝔖𝔱𝔢𝔞𝔪
𝔗𝔯𝔞𝔴𝔩𝔢𝔯𝔰 𝔏𝔬𝔯𝔡 𝔇𝔢𝔯𝔯𝔞𝔪𝔬𝔯𝔢 𝔞𝔫𝔡 𝔇𝔲𝔫𝔫𝔢𝔱𝔱, 1933.

NOW THE LABOURER'S TASK IS O'ER.

There the tears of earth are dried, There its hidden things are clear
There the work of life is tried by a Juster Judge than here.
Father in Thy gracious keeping, Leave we now Thy servants sleeping.

JOSEPH SUMMERS, Port Missionary

It was usually the job of the port missioner of the RNMDSF (Royal National Mission to Deep Sea Fishermen) or of the Fishermen's Bethel, to break the news of a man lost at sea. This card records a memorial service to the loss of *Lord Derramore* and the *Dunnett*, 5 March 1933, in the Barents Sea.

St Romanus (H223) built at Beverley in 1950. She was lost off Norway, probably on 11 January 1968, with the death of twenty men. Weather was extreme, with freezing fog and severe icing.

Kingston Peridot (H591) built at Beverley in 1948, she was lost while off Iceland, probably on the evening of 26 January 1968. No Mayday was heard and she is reckoned to have turned turtle, destabilised by a build-up of ice. All the crew were drowned.

Opposite below: The interior of the Fishermen's Bethel, on the Hessle Road, close to the Boulevard and at the heart of Hull's fishing community. The honours boards record the names of fishermen lost at sea.

Ross Cleveland (H61) built at Aberdeen in 1949 and lost off Iceland on 4 February 1968. She keeled over and sank in atrocious conditions, while the skipper was still talking on the radio. Three Hull trawlers and fifty-eight men were lost in less than a month; only one man, Harry Eddom, survived.

This photograph shows the build-up of ice on the wheelhouse of a trawler in northern waters in the winter. A crewman is at the winch, a crucial piece of machinery for controlling the trawl warps. If it fails, fishing is no longer possible.

The *Gaul* (H243) built in 1972 by Brook Marine of Lowestoft. On 8 February 1974, while commanded by Skipper Peter Nellist, she was lost in a gale off the North Cape of Norway with all thirty-six of her crew. No Mayday was transmitted and the circumstances of her loss are still being investigated.

In severe weather a lot of sea water washes on to the after deck and a stern trawler can be overwhelmed by the sheer weight of water. Even a relatively small amount reaching the factory deck can result in an excess of free surface water which can destabilise the vessel.

The stern trawler *Orsino* (H410). After the triple tragedy in 1968, a support or 'mother' ship was stationed on the Iceland grounds to give up-to-date meteorological reports and offer medical and other assistance. *Orsino* was the first vessel to take this role, sailing on 29 November 1968.

Eleven

The Stern Trawler

Fairtry, 280ft long. Her freezing equipment could handle thirty-five tons of cod fillets every twenty-four hours. Devised by Salvesens of Leith, based on their experience with whale factory ships, it was the first stern-fishing vessel with a ramp. She landed her first catch at Alexandra Dock, Hull, in 1955, being too big for the fish dock. She was the model for the Russian Puskin class factory ships.

Lord Nelson, built by Rickmers Werft of Bremerhaven for Associated Fisheries, she was the first stern trawler in the Hull fleet. Arriving 30 June 1961 she generated huge local interest. Stern fishing was the first fundamental advance in fishing methods since the introduction of steam eighty years before.

Above: Junella (H347) built 1962 by Hall Russell of Aberdeen for Marr's of Hull. She was a diesel-electric stern trawler, 240ft long and capable of 16 knots. Marr's Ltd had experimented with a freezer plant in the sidewinder trawler *Marbella*, a four-year programme which culminated in the *Junella*.

Right: Junella leaving the Hull fish dock, a tight fit in the lock pit! She sailed on her maiden voyage on 17 July 1962 for Newfoundland and, after a thirty-two-day trip, landed 5,500 kits of frozen fish, mainly cod.

Opposite below: Lord Nelson (H330). At 222ft long, she had two fish rooms: one storing fish in crushed ice in the old way; the other freezing the catch using plate freezers, with a capacity for twenty-five tons a day. Laid up in 1976 after the third Cod War, she was scrapped in 1981.

Junella in the Humber. Skipper Drever, the 1961 winner of the Silver Cod trophy, set a national record with the *Junella* of 420 tons of fish for a forty-two-day round trip to Newfoundland.

Most of the fish caught by the stern trawling fleet was frozen whole, and here we see a cargo being unloaded from the fish hold ready for transport to store or the processing plant.

St Finbarr (H308) built 1964 by Ferguson Bros of Port Glasgow for Thomas Hamlings of Hull; 211ft long. On her maiden voyage she set a new national record of 488 tons, 17cwt, for a thirty-seven-day trip to Labrador. She was lost after a fire in her accommodation area on 24 December 1966, twelve men died from a crew of twenty-five. The *Orsino* picked up survivors and attempted a tow, but the tow parted in heavy weather and *St Finbarr* sank.

C.S. Forester (H86): built at the Beverley shipyard in 1969, she was 186ft long and 768 gross tons. She was a 'fresher', the catch being preserved in crushed ice not frozen solid. Despite mechanical problems she was Champion British Trawler in 1976, 1977 and 1978 and winner of the Hull Challenge Shield. In 1978 a catch from Norway grossed £92,458. She was sold in 1980 to Iceland, after the Cod War.

Cape Kennedy (H353), a whole fish freezer weighing 2,271 tons, built by Cochranes of Selby in 1965 for Hudson Bros of Hull. In 1966 she was bought by Ross Group, renamed *Ross Intrepid* and, nine years later, became part of the B.U.T. fleet. She was laid up in 1975 as uneconomical and was sold to Norwegians in 1978 for conversion to a seismographic survey vessel.

His and Hers! Sir Fred Parkes (H385), built in 1966 by Hall Russell, these whole fish freezers – named after the owner of the firm and his wife, Sir Frederick Parkes and Lady Parkes – were the first stern trawlers in the Boston Deep Sea Fisheries fleet. In 1982 she ceased fishing and became a stand-by vessel for the oil industry.

The Lady Parkes (H397) made her first voyage in May 1966, two months after her sister ship *Sir Fred Parkes*. In 1968 she was the top British freezer with a catch totalling 3,790 tons, and in 1969 4,169 tons in seven trips totalling 288 days at sea. Sold to French owners in 1977 for geophysical survey work in the Antarctic.

Kurd (H242). Originally the *Ranger Callisto*, she was built by Brook Marine of Lowestoft in 1972 as a factory freezer ship for the Ranger Fishing Co., a subsidiary of P&O. Acquired by B.U.T. in 1973 for the Hull fleet, she was sold in 1981 to Norwegian owners for use as a diving support vessel.

Ranger Castor (SN18). Built by Brooke Marine in 1972, she was the fourth of the Ranger Fishing Co.'s 'C' class. The class was nicknamed the 'floating Hiltons' because of their superior crew accommodation, soap and bed linen provided, and films available off duty! Acquired by B.U.T. in 1973 and renamed *Gaul*, she was lost off Norway in 1974.

Farnella, built 1972, the first of the sister ships built by Swan Hunter (Clelands) at Wallsend for J. Marr & Son. She was never out of the top twenty British freezers. In 1982 she was employed as a minesweeper in the Falklands, was used for geophysical survey off the US coast in 1984, and in 1986 was chartered by the US Geophysical Survey Dept. for a period of five years.

Hammond Innes (H180) was built at Beverley shipyard and launched with a splash on 16 May 1972 for Newington Trawlers of Hull. Larger than the *C.S. Forester*, but of similar design, she was a 'fresher'.

Hammond Innes (H180). During a four-year fishing career she won the Hull Challenge Shield in 1974 (31,275 tons) and 1975 (36, 702 tons) and was the first trawler to break the £50,000 barrier for earnings from a single voyage. She was sold to Canada for conversion to a fisheries research vessel.

Sea Fridge Osprey (H137): launched at Aalesund, Norway on 4 March 1972 for Sea Fridge Ltd, she was a joint venture between Norwegian and Canadian interests, managed by Newingtons of Hull. After a short career she transferred to Norway in 1975 and was converted for seismographic survey in 1982.

Northella (H206) built by Swan Hunter (Clelands) at Wallsend and launched 2 July 1973 for J. Marr & Son. She came third in the Dolphin Bowl Competition in 1978, landing 4,579 tons. In 1982 she was employed minesweeping in the Falklands and was chartered by the navy for escort and training duties in 1983.

Junella (H294), 199ft long and 1,614 gross tons, was built by Swan Hunter (Clelands) at Wallsend in 1975. She took the name of Marr's pioneering freezer trawler. She landed 620 tons on her maiden voyage despite quota restrictions. During April-October 1982 she served as a minesweeper in the Falklands.

Arctic Buccaneer (H188) built by Stocznia Im Komuny Parskiej at Gdynia in 1973 as a whole fish freezer for the Boyd Line of Hull. At 281ft long and 1,660 gross tons, she was the biggest trawler in the British fleet, arriving in Hull on 6 December under command of Skipper Terry Thresh.

Arctic Buccaneer: this view of the bridge gives an impression of the scale of the vessel. She initially suffered severe propeller vibration and lost an entire year's fishing while this was corrected in Poland.

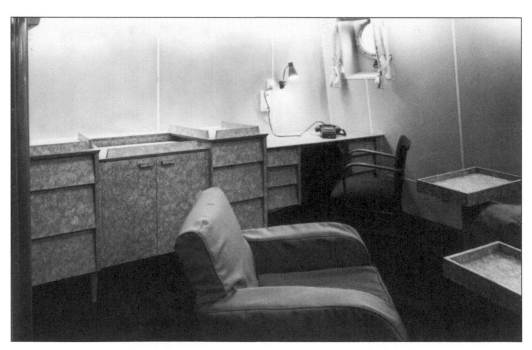

Arctic Buccaneer: the skipper's very spacious accommodation. She sailed on her maiden voyage on 17 April 1975 under Skipper Dicky Bryant, and returned from the White Sea with 724 tons of cod, haddock, plaice, coalfish and others. She won the Dolphin Bowl in 1978 and 1979.

Arctic Buccaneer: a clear view of the stern ramp and gantry for towing the net.

Arctic Buccaneer: a view aft towards the stern ramp – see the open hatches to the fish factory deck below.

Arctic Buccaneer: the vertical plate freezers capable of freezing sixty tons a day.

A full cod end: the trawl net has just been hauled up the stern ramp.

Twelve

Cod Wars and an Uncertain Future

Declaration of a 200-mile limit around the coast of Iceland in 1975 led to the third Cod War as British vessels continued to fish in disputed waters. Here the Icelandic fishery protection vessel *Tyr* manoeuvres close to a Hull trawler.

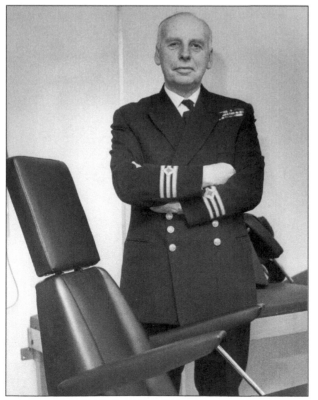

Above: In 1970 the four-masted schooner *Albatross*, a Swedish sail-training ship, was bought and converted as a mother ship for the trawler fleet in Icelandic waters. Renamed *Miranda* she was on duty in 1972 when the declaration of a fifty-mile limit sparked the second Cod War.

Left: Dr Robert W. Scott of Harrogate, medical officer aboard the *Miranda*. She had a merchant navy crew, a fully equipped hospital and gave regular met. reports to the fishing fleet. Scott was M.O. during the second and third Cod Wars (*Miranda* was sold in 1980).

While fishing at night, under lights, the Icelandic fisheries protection vessel *Tyr* attempts to draw its cutting gear across the trawlers warps as she tows a net.

Increasing fishing restraints coincided with the OPEC oil cartel pushing up the price of fuel and this resulted in a wholesale disposal of the 'side-fishing' fleet. Many were sold abroad or converted for oil rig support; a lot were scrapped like these in Drapers Yard, Victoria Dock, Hull, November 1975.

The imposition of quotas and limits was necessitated by falling fish stocks and trophies and league tables, which encouraged fishing to the physical limits of the trawlers, were abandoned. The Silver Cod was last awarded in 1968; on the left is the trophy and on the right the miniature kept by the winning skipper. The Distant Water Challenge Shield was last awarded in 1977.

Hull's fishing was in the doldrums after the last Cod War but a rebuilding of confidence resulted in additions to the fleet. The *Thornella*, launched at Selby in 1988, was the first new trawler in the Marr fleet for twelve years.

The *Arctic Corsair* (H320) of Boyd Line, Hull. A reminder of the heyday of post-war fishing, this side-winder is preserved as a museum ship. Used for mackerel fishing in 1980-1981 she was laid up but resumed fishing in 1985 and made several voyages grossing £80,000 to £90,000 each, a swan song to side-fishing out of Hull.

Cordella (H177). Marr's have been successful at diversification, with charters for survey work and an unexpected bonus in 1982 with the need for minesweepers in the Falklands. *Cordella* was one of several Marr trawlers sent to the South Atlantic.

Criscilla: a Marr trawler which found a role as a Fishery Patrol vessel.

Thornella (H96) re-equipping in Hull docks, 19 October 2002. Built in 1988, she and her sister ship *Lancella* (H98) remain in the fleet. In January 1990 *Lancella* achieved a British record, grossing £260, 324 for a single trip. The state of fishing stocks world-wide is the major problem now facing all fishing fleets.